M000189505

Vine Pruning

By Frederic T. Bioletti

UNIVERSITY OF CALIFORNIA.

AGRICULTURAL EXPERIMENT STATION.

BERKELEY, CAL.

E. W. HILGARD, Director.   BULLETIN No. 119.

DECEMBER, 1897.

VINE PRUNING.

By F. T. BIOLETTI.

The literature relating to the pruning and training of the vine is already very voluminous, but there seems to be no one work which treats the subject in a thorough and convenient way for California vine-growers. Publications in English refer generally to methods suited to the Eastern States or to hot-house cultivation, while foreign publications, besides being more or less inaccessible, treat the subject so widely that the grower is at a loss what to choose from such a mass of material. It is the purpose of this Bulletin, therefore, to present a brief summary of what in foreign methods seems useful and applicable to California conditions, together with the results of experiments on the University of California vine plots, and of observations made in numerous vineyards in various regions of the State.

Almost every vine-growing district has its peculiar systems of training, ranging from the non-training usual in parts of Italy, where the vine spreads almost at will over trees planted for the purpose, to the acme of mutilation practiced in many localities where the vine is reduced to a mere stump barely rising above the surface of the ground. These various systems will not be discussed here, but only those which experience has shown to be most adapted to California conditions.

No account, however detailed, of any system can replace the intelligence of the cultivator. For this reason the general principles of plant physiology which underlie all proper pruning and training are discussed in connection with the several systems described. This should aid the grower in choosing that system most suited to the conditions of his vineyard, and to modify it to suit special conditions and seasons. All the operations of pruning, tying, staking, etc., to which a cultivated vine owes its form, are conveniently considered together.

No cultivated plant is susceptible of such a variety of modes of training as the vine, and none can withstand such an amount of abuse in this matter and such radical interference with its natural mode of growth. On the other hand, no other plant, perhaps, is so sensitive to proper treatment, or responds so readily to a rational mode of pruning and training.

OBJECTS OF PRUNING.—The objects of pruning are (a) to facilitate cultivation and gathering, (b) to increase the average yield, and (c) to improve the quality of fruit. The vine must not be trained so high that the grapes are difficult to gather, nor allowed to spread its arms so wide that the cultivation of the ground is unduly interfered with. Vines untouched by the pruner's knife bear irregularly; a year of over-bearing being followed by several of under-bearing as a consequence of exhaustion caused by a too severe drain on the reserve forces of the plant. The grapes on untrained or improperly trained vines are exposed to different conditions of heat and light, and consequently develop and ripen unevenly.

PHYSIOLOGICAL PRINCIPLES.—The main facts regarding the physiology of the vine to be kept in mind in this connection are:

1. The vine feeds by means of the green coloring matter (chlorophyll) of its leaves. It obtains the sugar, starch, etc., which it needs from the carbonic acid of the air which is converted into these substances by the chlorophyll under the influence of light. A certain amount of green leaf surface functioning for a certain time is necessary to produce sufficient nourishment for the vital needs of the vine and for the production of a crop. Those leaves most exposed to the direct rays of the sun are most active in absorbing food. The youngest leaves take all their nourishment from the older parts of the plant: somewhat older leaves use up more nutrient material in growing than they absorb from the air. A young shoot may thus be looked upon as, in a sense, parasitic upon the rest of the vine. The true feeders of the vine and of its crop are the *mature, dark-green leaves.*

2. Within certain limits the fruitfulness of a vine or of a part of a vine is inversely proportional to its vegetative vigor. Methods which tend to increase the vegetative vigor of a vine or of a part of a vine tend to diminish its bearing qualities, while, on the contrary, anything which diminishes vegetative vigor tends to increase fruitfulness. Failure to reckon with this fact and to maintain a proper mean between the two extremes leads, on the one hand, to comparative sterility, and, on the other, to over-bearing and premature exhaustion.

3. The vine tends to force out terminal buds and to expend most of its energy on the shoots farthest from the trunk. To keep the vine within practical limits, this tendency must be controlled by the removal of the terminal buds, or by measures which check the flow of sap and force the growth of buds nearer the stock.

4. The nearer a shoot approaches the vertical the more vigorous it will be.

5. The size of shoots and of fruit is, within certain limits, inversely as their amount. That is, with a given vine, or arm of a vine, the fewer shoots allowed to grow the larger each will be, and the same is true of bunches of fruit.

6. Other conditions being equal, an excess of foliage is accompanied by a small amount of fruit; an excess of fruit by diminished foliage.

7. Shoots coming from one-year-old wood growing out of two-year-old wood are alone to be depended on for fruit. Other shoots are usually sterile.

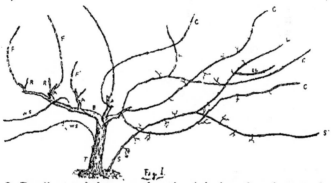

Fig. I

8. Bending, twisting or otherwise injuring the tissues of the vine or its branches tend to diminish its vegetative vigor, and therefore, unless excessive, to increase its fertility.

A description of a typical vine giving the names of the principal parts, will make clear the accounts of methods to be given later. Fig. I represents a vine of no particular order of pruning, showing the various parts. The main body of the vine (T) is called the trunk or stem; the principal divisions (B) branches; the smaller divisions (A) arms, and the ultimate ramifications (C) shoots when green, and canes when mature. A shoot growing out of the vine above ground on any part older than one year (WS) is called a water sprout. Shoots coming from any part of the vine below ground (S) are called suckers. When a cane is cut back to 1, 2, 3, or 4 eyes it is called a spur (R).

When a shoot or cane of one season sends out a secondary shoot the same season, the latter (L) is called a lateral.

Fig. II represents an arm of a vine as it appears in winter after the leaves have fallen. The canes (W1) are the matured shoots of the previous spring. W2, W3, W4 represent 2, 3, and 4-year-old wood respectively. Near the base of each cane is a basal bud or eye (B°). In counting the number of eyes on a spur the basal eye is not included. A cane cut at K1 for instance leaves a spur of one eye, at K2 a spur of two eyes and so on. When more than four eyes are left the piece is generally called a fruiting cane (Fig. I, F). The canes (C,C1) coming from two-year old wood (W2) possess fruit buds; that is, they are capable of producing fruit-bearing shoots. Water sprouts (WS) and suckers (S) do not ordinarily produce fruit-bearing shoots. Below the basal bud each cane has one or more dormant buds (b Fig. III) which do not grow unless the number of eyes left by pruning or frost is insufficient to relieve

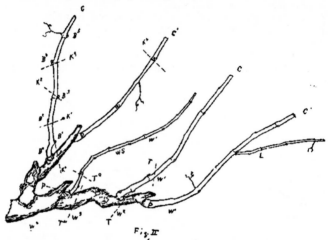

Fig. II

the excess of sap pressure. These buds produce sterile shoots. Each eye on a cane has, at its base, two dormant buds. One of these sometimes grows out the year it is formed, making a lateral (L, Figs. I, II). These laterals may send out secondary laterals (SL, Fig. I). It is on the laterals and secondary laterals that the so-called second and third crops are borne.

PRUNING FOR WOOD AND FOR FRUIT.—One of the chief aims of pruning is to maintain a just equilibrium between vegetative vigor and fertility. We must, then, prune for both wood and fruit. A vine which has become enfeebled by over-bearing should be pruned for wood. By this is meant that only a small number of buds should be left. As all the energies of the vine have to be expended on a small number of shoots, these shoots grow with more than ordinary vigor. Under these conditions the vine bears little; first, because the eyes near the bases of the canes, which are the only ones left in very short pruning, are naturally less fruitful than those farther removed from the main body of the vine; and second, because an exceptionally vigorous shoot is generally sterile. The vine is thus strengthened, and, as the stores of nutriment provided by a vigorous vegetation are not drawn upon by a heavy crop, the increased vigor of the vine is more marked the second year. The second year, therefore, more wood may be left and the crop increased without detriment to the vine.

. On the other hand, a vine which "goes to wood" must be pruned for fruit. For this purpose we increase the number of buds left and choose the most fruitful wood. The largest canes are the least fruitful, while the smallest have not the necessary vigor to support a large crop. The best cane to leave for fruit then is one of medium size, with well-formed eyes.

PROPER METHOD OF MAKING CUTS.—It is by no means a matter of indifference just where the cut is made in removing a cane or arm. This will be made clearer by referring to Fig. III. The upper part of the spur is represented as split in two longitudinally in order to show the internal structure of the cane. It will be noted that at each bud there is a slight swelling of the cane. This is called a node, and the space between an internode. The internodes are filled with soft pith, but at each node there is a growth of hard wood extending through the cane. Now, if the cane be cut off at C1, in the middle of an internode, the pith will shrink away and leave a little hollow in which the rain collects. This is an excellent breeding place for fungi and bacteria, which cause rotting of the pith and

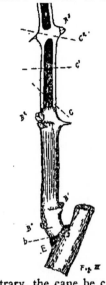

frequently kill the bud. If, on the contrary, the cane be cut at C2, through a node, a protecting cover of hard wood is left, which is an effectual barrier against decay organisms. If a spur projects too far from the vine and it is desirable to make it as short as possible in order not to interfere with cultivation, it should be cut at C and the cut made as nearly vertical as possible. This allows the water to run off, and leaves less pith to foster the growth of the fungi. At the base of the cane there is a slight enlargement (E). In removing a cane completely the cut should be made just above this enlargement. This is the most favorable place for healing, as it makes the smallest possible wound and does not leave a projecting stump of dead wood to prevent the healing tissues from closing over the wound. In removing a piece of older wood, as at K° and T1, Fig. II, it is advisable not to cut too close for fear of injuring the spur by the drying out of the wood. The projecting pieces of dead wood left in this way should be carefully removed the next year in order to allow the wound to heal over. The large cuts which are thus occasionally necessary are most easily performed by means of a well-made and well-sharpened pair of two-hand pruning shears. These shears are often to be preferred to the ordinary one-hand shears because they render the cutting through the nodes easier and do away almost entirely with the necessity of a saw. Of course, a careless workman may split and injure vines seriously by using long-handled shears clumsily, but the bending of arms to facilitate cutting with the one-hand shears often results in the same evil. The one-hand shears, however, are more convenient when many long fruiting canes are left, as the necessary trimming off of tendrils and laterals is more easily performed with them.

SHORT AND LONG PRUNING. — The winter-pruning of the vine consists in cutting off a certain amount of the mature wood of the immediately preceding season's growth (canes), and occasionally of the older wood. The main problem of winter-pruning, then, resolves itself into determining how much and what wood shall be left. In all kinds of pruning most of the canes are removed entirely. In *short-pruning* the remainder are cut back to spurs of one, two or three eyes. The number of spurs is regulated by the vigor and age of the vine. This mode of pruning can be used only for varieties in which the eyes near the base of the cane are fruitful. For all other cases long or half-long pruning is necessary.

In *half-long pruning* certain canes are left with from four to·six eyes, according to the length of the internodes. These canes or fruit-spurs will bear more fruit than short spurs for three reasons: 1, because there will be more fruit-bearing shoots; 2, because the upper eyes are more fruitful than the lower; and 3, because a larger number of eyes being supplied with sap from the same arm, each shoot will be less vigorous and therefore more fruitful. Owing, however, to the tendency of the vine to expend the principal part of its vigor on the shoots farthest removed from the base of the canes, the lower eyes on the long spurs will generally produce very feeble shoots. In order, then, to obtain spurs of sufficient vigor for the next year's crop it would be necessary to choose them near the ends of the long spurs of the previous year, if no others were left. This would result in a rapid and inconvenient elongation of the arms. In order to avoid this it is necessary to leave a spur of one or two eyes below each long fruiting spur, that is to say, nearer the trunk. These short wood spurs having only one or two eyes, will produce vigorous canes for the following year, and the spurs which have borne fruit may be removed altogether, thus preventing an undue elongation of the arms. In half-long pruning, however, it is very hard to retain the proper equilibrium between vigor and fruitfulness. If a little too much wood is left the shoots from the wood spurs will not develop sufficiently,and the next year we have to choose between leaving small under-sized spurs near the trunk and spurs of proper size too far removed from the trunk. In long pruning this difficulty, as will be seen, is to a great extent avoided.

In *long pruning* the fruit spurs of half-long pruning are replaced by long fruit canes. These are left two or three feet long, or longer. The danger here that the vine will expend all its energies on the terminal buds of these long canes and leave the eyes of the wood spurs undeveloped is still greater than in half-long pruning. This difficulty is overcome by bending or twisting the fruit canes in some manner. This bending causes a certain amount of injury to the tissues of the canes, which tends to check the flow of sap towards their ends. The sap pressure thus increases in the lower buds and forces them out into strong shoots to be used for spurs for the next pruning. The bending has the further effect of diminishing the vigor of the shoots on the fruit canes and thus increasing their fruitfulness.

This principle of increase of fruitfulness by mechanical injury is very useful if properly understood and applied. It is a well-known fact that vines attacked by phylloxera or root rot will for one year bear an exceptionally large crop on account of the diminution of vigor caused by the injury to their roots. A vine also which has been mutilated by the removal of several large arms will often produce heavily the following year. In all these cases, however, the transient gain is more than counter-balanced by the permanent injury and loss. The proper application of the principle is to injure tissues only of those parts of the plant which it is intended to remove the next year (fruit canes), and thus increase fruitfulness without doing any permanent injury to the plant.

PRUNING OF YOUNG VINES.—When a rooted vine is first planted, it should be cut back to two eyes. If the growth is not very good the first season, all the canes but one should be removed at the first pruning, and that one left with two or three eyes, according to its strength. The next year, or the same year in the case of strong growing vines in rich soil, the strongest cane should be left about 12 inches long and tied up to the stake The next year two spurs may be left, of two or three eyes each. These spurs will determine the position of the head or place from which the arms of the vine spring. It is important, therefore, that they should be chosen at the right height from the ground. From ten to twenty inches is about the right height; the lowest for dry hillsides where there is no danger of frost; the highest for rich bottom lands where the vine will naturally grow large. Vines grown without stakes will have to be headed lower than this in order to make them support themselves. In the following few years the number of spurs should be increased gradually, care being taken to shape the vine properly and to maintain an equal balance of the arms.

In general, young vines are more vigorous than old, and tend more to send out shoots from basal and dormant buds. They should, therefore, be given more and longer spurs in proportion than older vines. They also tend to bud out very early in the spring, and are thus liable to be frost-bitten. For this reason they are generally pruned late (March) in frosty locations. This protects them in two ways. In the first place, in unpruned vines the buds near the ends of the canes start first and relieve the sap pressure, and though these are caught by the frost the buds near the base, not having started, are saved. In the second place, the pruning being done when the sap is flowing there is a loss of sap from the cut ends of the spurs which further relieves the sap pressure and retards the starting of the lower eyes. This method of preventing the injury of spring frosts by very late pruning has been tried with bearing vines, but is very injurious. Older vines being less vigorous are unable to withstand the heavy drain caused by the profuse bleeding which ensues; and though no apparent damage may be done the first year, if the treatment is continued they may be completely ruined in three or four years.

SYSTEMS OF PRUNING.

The systems of pruning adapted to vineyards in California may be divided into six types according to the form given to the main body of the vine and the length of the spurs and fruiting canes

A. Vine pruned to a head, with short arms.

    I. With spurs of two or three eyes only (short pruning).

    II. With wood spurs of one or two eyes and fruit spurs of four. to six eyes (half-long pruning).

    III. With wood spurs of one or two eyes and long fruit canes, (long-pruning).

B. Vine with a long horizontal branch or continuation of the trunk.

    IV. With spurs of two or three eyes only (short pruning).

    V. With wood spurs of one or two eyes and fruit spurs of four to six eyes (half-long pruning).

    VI. With wood spurs of one or two eyes and long fruit canes (long-pruning).

These types are applicable to different varieties of vines according —(1) To the natural stature of the vine—that is to say, whether it is a large or small grower and tends to make a large, extended trunk or a limited one.—(2) To the position of the fruit buds. In some varieties all the buds of the canes are capable of producing fruitful shoots, while in others the one, two or three buds nearest the base produce only sterile shoots.—(3) To the size of the individual bunches. It is necessary in order to obtain a full crop from a variety with small bunches to leave a larger number of eyes than is necessary in the case of varieties with large bunches.

What type or modification of a type shall be adopted in a particular instance depends both on the variety of vine and on the nature of the vineyard. A vine growing on a dry hillside must not be pruned the same as another vine of the same variety growing on rich bottom land. In general, vines on rich soil, where they tend to grow large and develop abundant vegetation, should be given plenty of room and allowed to spread themselves, and should be given plenty of fruiting buds in order to control their too strong inclination to "go to wood." Vines on poor soil, on the contrary, should be planted closer together and pruned shorter, or with fewer fruiting buds, in order to maintain their vigor.

Type I.—This is the ordinary short pruning practiced in 90 per cent of the vineyards of California, and is the simplest and least expensive manner of pruning the vine. It is, however, suited only to vines of small growth, which produce fruitful shoots from the lowest buds, and of which the bunches are large enough to admit of a full crop from the small number of buds which are left by this method. The chief objection to this method for heavily bearing vines is that the bunches are massed together in a way that favors rotting of the grapes and exposes the different bunches unequally to light and heat.

Fig. IV. represents the simplest form of this style of pruning. The vine should be given, as nearly as possible, the form of a goblet, slightly flattened in the direction of the rows. It is important that the vine be kept regular and with equally balanced arms. This is the chief difficulty of the method and calls for the exercise of some judgment. From the first, the required form of the vine should be kept in view. On varieties with a trailing habit of growth vertical spurs must be chosen, and with some upright growers it will be found necessary to choose spurs nearer the horizontal.

The arms must be kept short for convenience of cultivation and

to give them the requisite strength to support their crop without bending or breaking. For this reason the lowest of the two or three canes coming from last year's spur should be left. For instance, on Fig. II the cane should be cut at K_2 or K_3, according as two or three eyes are needed, and the rest of the arm removed at $K°$. As even with the greatest care some arms will become too long or project in wrong directions, it is necessary to renew them by means of canes from the old wood or water sprouts. For instance, if the other arm represented on Fig. II were too long, it should be removed and replaced by another developed from the cane (WS). As the cane comes from three-year-old wood it cannot be depended on to produce grapes. For this reason it is best the first year to prune the arm at T, leaving a spur for fruit, and cut the water sprout at $T°$ leaving a wood spur of one eye. The next year the cane coming from the first eye of WS can be left for a fruit spur, and the arm removed at T_1. The cutting back of an elongated arm should not be deferred too long, as the removal of old arms leaves large wounds which weaken the vine and render it liable to attacks of fungi.

In order to maintain the equilibrium of the arms it is often necessary to prune back the more vigorous arms severely in order to throw the strength of the vine into the weaker arms. If the vine appears too vigorous, that is if it appears to be "going to wood" at the expense of the crop, two spurs may be left on some or all of the arms. In this case the upper spur should be cut above the third eye (K_4 Fig. 11), and the lower above the first or second (K_1 or K_2). This will cause the bulk of the fruit to be borne on the upper spur, and the most vigorous shoots to be developed on the lower, which provides the wood for the following year. This is an approach to the next (half-long) method of pruning.

Type II.—Vines which require more wood than can well be given by ordinary short pruning, or of which the lower eyes are not sufficiently productive, may in some cases be pruned in the manner illustrated by Fig. V. For some varieties it is necessary to leave spurs of only three eyes, as at S; for others, short canes of four or five eyes must be left, as at CC. These shorter spurs can be left without support, but the longer ones require some arrangement to prevent their bending over with the weight of fruit and destroying the shape of the vine. In some cases simply tying the ends of the canes together will support them fairly well, but it is better to attach them to a stake and bend them at the base a little when possible in order to retard the flow of sap to the ends. It is very necessary to leave strong spurs of one eye (not counting the basal eye) in order to provide wood for the following year. At the pruning following the one represented in the cut the fruiting part of the arms will be removed at KK and a new fruiting spur or cane made of the cane which comes from the eye on the wood spurs W. The basal bud on W will in all probability have produced a cane which can be cut back to one eye to furnish a new wood spur. If this is not the case it shows that too much wood was left the first year, and therefore no fruit cane should be left on this arm, but only a single spur of two or three eyes. This will be a return to short pruning, and must be resorted to whenever the small size of the canes or the failure to produce replacing wood near the head of the vine shows that the vigor is diminishing. If, on the contrary, the arm shows an abundance of vigorous canes, proving that the vine has not overborne, a fruit cane may be left from one of the shoots coming from the lower buds of the fruit cane C, and a new wood spur of two eyes left on the shoot coming from the wood spur of the previous year (W). In this case, the removal of the arm at K is deferred one year, and the extra vigor of the vine is made use of to produce an extra crop.

Type III.—This style is an extension of the principles used in type II, as will be understood by referring to Fig. VI. The fruiting canes are left still longer, and in some cases almost the full length of the cane. As each cane will thus produce a large amount of fruit, fewer arms are necessary than in the preceding method. It is especially necessary to leave good, strong spurs of one or two eyes to produce wood for the following year. There are various methods

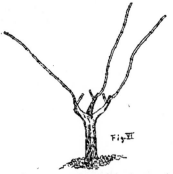

Fig. VI

of disposing of the long fruiting canes, the worst of which is to tie them straight up to the stake, as was recommended for the half-long canes. In the latter case, owing to their shortness, a certain amount of bending of the canes is possible with this method of tying. With long canes, on the contrary, it usually allows of no bending, and as a result there ensues a vigorous growth of shoots at the ends of the fruiting canes, and little or no growth in the parts where it is necessary to look for wood for the following year. Often, indeed, each long cane will produce only three shoots and these from the three terminal eyes, all the other eyes of the cane remaining dormant. The object of long pruning is thus doubly defeated, 1st because no more shoots are produced than by short pruning, and 2nd, because the shoots which should produce fruit are rendered especially vigorous by their terminal and vertical position, and therefore less fruitful. Each year all this vigorous growth of wood at the ends of the canes must be cut away in order to keep the vine within practical bounds, and the fruit canes renewed from the less vigorous cane below. These canes are less vigorous because the main strength of the vine has been expended on the upper canes which are most favorably placed for vegetative vigor. Vines treated in this way may be gradually exhausted though producing only a moderate or small crop of fruit, by being forced to produce an abundant crop of wood.

One of the simplest ways of tying the fruiting canes is illustrated by Fig. VII. The canes are bent into a circle, the ends tied to the stake near the head of the vine, and the middle of the circle attached higher up. The tying should be done so that the cane receives a severe bend near the base—that is about the region of the second and third eyes. This can usually be accomplished by tying the end of the cane first, and then pressing down on the middle

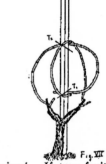

Fig. VII

of the bow until the desired bend is attained. If two fruiting canes are left, they should be made to cross each other at right angles in order to distribute the fruit as equally as possible. As a rule more than two canes should not be tied up in this way as it makes too dense a shade and masses the fruit too much.

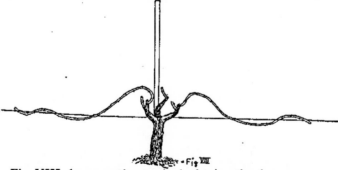

Fig. VIII

Fig. VIII shows another method of tying the long canes. A horizontal wire is stretched along the row at about fifteen to twenty inches above the ground. To this the fruiting canes should be attached, using the same precaution of bending the canes near the bases. The upper part of the canes is not bent in this case as in the last, but the necessary diminution of vigor and increase of fruitfulness is brought about by the horizontal position. Two canes may be attached to the wire on each side. The stake is best used to support the shoots destined for the wood for the following year. This makes it possible, where topping is practiced, to cut off the ends of the shoots from the fruiting canes and to leave the rest their full length. Another or even two other wires may be used above the first for more canes, but this is seldom profitable, and considerably increases the cost both of installation and or pruning.

This style of pruning is especially favorable to varieties of small growth which bear small bunches and principally on the upper eyes, and to varieties of larger growth in hilly or poor soils. One of its main objections is that it renders some varieties more liable to sunburn.

It will be noticed that the long-pruned vines are represented in the figures as having much fewer arms than the short-pruned. This is necessary and important. In order to maintain a well-balanced vine and keep it under control, there should be only about as many arms as long canes, or at most one or two more.

Types IV, V and VI.—The three styles of pruning so far described have been fairly thoroughly tested in California, and each has been found applicable to certain varieties and conditions. There are some varieties, however, which do not give good results with any of these systems. This is the case with many valuable table grapes, especially when grown in rich valley soil, where they should do best. For these cases some modification of the French cordon system is to be recommended. Little trial of this method has been made as yet, but what has been done is very promising. The tendency of many grapes to coulure is overcome, and rich soils are made to produce crops in proportion to their richness. The method consists essentially in allowing the vine to grow in a more or less horizontal direction for several feet, thus giving a larger body and fruiting surface.

The treatment of the young vines the first year is the same as for

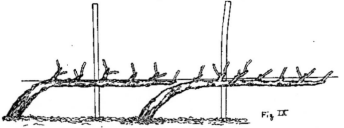

Fig. IX

head pruning, as already described. As soon as the young vine produces a good, strong shoot it is tied up to the wire and to the stake which is placed between the vines in the rows. Each vine should finally reach its neighbor, but it requires two or three years for this if the vines are six or seven feet apart in the rows. It is possible by cutting the vine back nearly to the ground for the first year or two to obtain a cane which will stretch the whole distance between the vines at the first tying up; but this is not necessary nor advisable. Neither is it advisable to make a very sharp angle (almost a right angle) as is usually done in regular cordon pruning, on account of the difficulty of preventing the vine from sending out an inconvenient number of shoots at the bend. The vine might be grown with two branches, one stretching in either direction, but this has been found inconvenient on account of the difficulty of preserving an equal balance of the branches. The direction in which the vine is trained should be that of the prevailing high winds, as this will minimize the chances of shoots being blown off. When the cordon or body of the vine is well-formed, it may be pruned with all the modifications of short, half-long and long pruning already described in head pruning, and the same precautions are necessary to preserve the balance and symmetry of the vine and to maintain it at the highest degree of fruitfulness without unduly exhausting it.

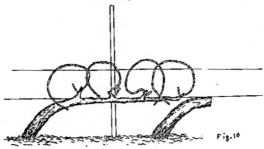

Figs. IX and X will sufficiently illustrate the way of shaping and tying short and long-pruned vines. For some table grapes extension of the method shown in Fig. IX in the direction of half-long pruning is useful. On a heavy soil the short spurs do not provide sufficient outlet for the vigor of the vine, while long pruning would unduly increase the number of bunches on a single cane, and so reduce their size, which would deteriorate from their value as table grapes.

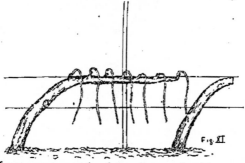

Fig. XI represents a style of pruning used with success in some of the richest low-lying soils of France. The body of the vine is raised up to a height of two and a half or three feet above the soil, a useful means of lessening the danger from spring frosts. The fruit canes are bent vertically downward thus restricting the flow of sap sufficiently to force out the lower buds of the fruit canes into strong shoots which can be used for fruit canes of the following year. This does away, to some extent, with the necessity of leaving wood spurs, and much simplifies the pruning. Arms, of course, are formed in time, and very gradually elongate, so that it is necessary to remove one occasionally and replace it by a water sprout, as already explained under short pruning.

SUMMER PRUNING.

Some form of summer or green pruning is practiced in most California vineyards, if in the term we include all the operations to which the green shoots are subjected. There seems, however, to be little system used, and very little understanding of its true nature and object. In general, it may be said that green pruning of the vine is least needed and often harmful in warm, dry locations and seasons, and of most use under cool and damp conditions.

The principal kinds of green pruning are: 1. pinching; 2. suckering and sprouting; 3. topping; 4. removal of leaves.

Pinching consists in removing the extreme growing tip of a young shoot. It is necessary to remove only about half an inch to accomplish the purpose of preventing further elongation of the shoot as all growth in length takes place at the extreme tip. The immediate result of pinching is to concentrate the sap in the leaves and blossoms of the shoot, and finally to force out the dormant buds in the axils of the leaves. It has been found useful in some cases to combat coulure or dropping with heavy-growing varieties, such as the Clairette Blanche. It is also of use in preventing unsupported shoots from becoming too long while still tender, and being broken off by the wind. It can, of course, be used only on fruiting shoots and not on shoots intended for wood for the following year.

Suckering is the removal of shoots that have their origin below or near the surface of the ground. The shoots should be removed as thoroughly as possible, the enlargement at the base being cut off in order to destroy the dormant basal buds. An abundant growth of suckers indicates either careless suckering of former years, (which has allowed a mass of buds below the ground, a kind of subterranean arm, to develop, or too limited an outlet for the sap. The latter may be due to frost or other injuries to the upper part of the vine, but is commonly caused by too close pruning.

Sprouting is the removal of sterile shoots or "water-sprouts" from the upper part of the vine. Under nearly all circumstances this is an unnecessary and often a harmful operation, especially in warm, dry locations. An exception may perhaps be made under some conditions of varieties like the Muscat of Alexandria, which has a strong tendency to produce "water-sprouts" which, growing through the bunches, injure them for table and drying purposes.

Water-sprouts are produced from dormant buds in the old wood, and as these buds require a higher sap pressure to cause them to start than do the fruitful buds, the occurrence of many water-sprouts indicates that too limited a number of fruitful buds has been left upon the vine to utilize all the sap pumped up by the roots. To remove these water-sprouts, therefore, while they are young is simply to shut off an outlet for the superabundant sap and thus to injure the vine by interfering with the water equilibrium, or to cause it to force out new water-sprouts in other places. Any vigorous vine will produce a certain number of water-sprouts, but they should not be looked upon as utterly useless and harmful because they produce no grapes. On the contrary, if not too numerous, they are of positive advantage to the vine, being so much increase to the feeding surface of green leaves. Water-sprouts should be removed completely during the winter pruning, and the production of too many the next year prevented by a more liberal allowance of bearing wood.

Topping, or cutting off the ends of shoots, is done by means of a sickle or long knife. At least two or three leaves should be left beyond the last bunch of grapes. The time at which the topping is done is very important. When the object is simply to prevent the breaking of the heavy, succulent canes of some varieties by the wind, or to facilitate cultivation, it must of course be done early, and is well replaced by early pinching. These objects are, however, better attained by appropriate methods of planting and training. Early topping is inadvisable because it induces a vigorous growth of laterals which make too dense a shade, and it may even force the main eyes to sprout, and thus injure the wood for the next year. The legitimate function of topping is to direct the flow of food material in the vine first into the fruit, and second into the buds for the growth of the following year. If the topping is done while the vine is in active growth, this object is not attained; one growing tip is simply replaced by several. In this way, in rich, moist soils vines are often, by repeated toppings, kept in a continual state of production of new shoots, and as these new shoots consume more food than they produce, the crop suffers. Not only does the crop of the current year suffer, but still more the crop of the following year, for the vine devotes its energy to producing new shoots in the autumn instead of storing up reserve food-matrial for the next spring growth. If, on the other hand, the topping is done after all leaf growth is over for the season, the only effect is to deprive the vine of so much food-absorbing surface.

The topping, then, should be so timed that, while a further lengthening of the main shoot is prevented, no excessive sprouting of new laterals is produced. The exact time differs for locality, season and variety, and must be left to the experience and judgment of the individual grower.

Removal of Leaves.—In order to allow the sun to penetrate to and aid the ripening of late grapes it is often advisable late in the season to lessen the leafy shade of the vine. This should be done by removing the leaves from the center of the vines and not by cutting away the canes. In this way only those leaves are removed which are injurious, and as much leaf surface as possible is left to perform the autumn duty of laying up food-material for the spring. The removal of leaves should not be excessive, and if considerable, should be gradual, otherwise there is danger of sunburn. It is best, first, to remove the leaves *from below* the fruit. This allows free circulation of the air and penetration of the sun's rays which warm the soil and are reflected upon the fruit. This is generally sufficient, and in any case only the leaves in the center of the vine, and especially those which are beginning to turn yellow should be removed.

In the list of varieties which follows, an attempt has been made to indicate the mode of pruning which is likely, in the light of our pres· ent knowledge, to give the best results for each variety. It should be understood, however, that it is to some extent tentative and provisional. Many of the varieties have proved successful in certain soils and locations when pruned in the way indicated, but others have never, so far as we know, been tested in the way proposed. As these latter, however, have proved more or less unsuccessful under the common methods of treatment the method proposed is the one which seems most suitable to their habit and general characters. It seems probable that the tendency to coulure of some varieties such as the Muscat, Malbeck, Merlot, Clairette, etc., can be combatted to a great extent by appropriate methods of pruning and training. Unevenness of ripening and liability to sunburn of Tokay, Zinfandel, etc., can doubtless be controlled by the same means.

Very few varieties succeed under strictly short pruning, that is cutting back to one and two eyes, so that for most of the varieties in the first category the modification of short pruning which gives fruit spurs of three or four eyes and wood spurs of one eye is recommended.

Type I. Charbono, Cinsaut, Mataro, Carignane, Grenache, Petit and Alicante Bouschet, Aramon, Mourastel, Verdal, Ugni-blanc, Folle blanche, Burger, Zinfandel, Grüner Velteliner, Pevereila, Zierfahndler (?), Rother Steinschiller (on poor soils), Slankamenka, Green Hungarian (on poor soils), Blue Portuguese (on poor soils), Tinta Amarella, Moscatello fino, Pedro Ximenes, Palomino, Beba (?), Peruno, Mantuo, Mourisco branco, Malmsey, Mourisco preto, Feher Szagos, Muscat of Alexandria, Sultanina, Sultana, Barbarossa.

Type II. St. Macaire, Beclan (longer or shorter according to rich-- ness of soil), Teinturier male, Mondeuse, Marsanne, Chasselas, Muscatel, Grosse Blaue, Sauvignon blanc, Sauvignon vert, Nebbiolo, Fresa, Aleatico.

Type III. Cabernet Sauvignon and Cabernet Franc (on poor soils and hillsides), Verdot, Tannat, Gamai Teinturier, Gros Mansenc, Pinots, Meunier, Gamais, Pinot blanc, Pinot Chardonay, Rulander, Affenthaler, Johannisberger, Franken Riesling (on hillsides), Kleinberger, Traminer, Walschriesling, Rothgipfler, Lagrain (? perhaps short), Marzemino, Blue Portuguese (on rich soils), Barbera, Moretto, Refosco, Tinta de Madeira, Tinta Cao, Verdelho, Boal..

Type IV. Green Hungarian, Rother Steinschiller (on rich soils), Neiretta, Mission, West's Prolific, Robin noir.

Type V. St. Macaire and Mondeuse (on rich bottom soils), Tinta Valdepeñas, Marsanne, Clairette blanche, Semillon, Sauvignon blanc (on rich soils), Muscadelle du Bordelais, Vernaccia bianca, Furmint Bakator, Tadone, Gros Colman, Black Morocco (?), Cornichon (?), Emperor, Tokay (?), Almeria, Pizzutello, California black Malvoisie.

Type VI. Malbec, Petite Sirah and Serine, Cabernet Sauvignon and Cabernet Franc (on rich bottom soils), Merlot, Gros Mansenc (? on rich bottom soils), Chauché noir, Bastardo, Trousseau, Ploussard, Etraie de l'Adhui, Chauché gris, Franken Riesling (on rich soils).

CPSIA information can be obtained at www.ICGtesting.com
Printed in the USA
BVOW041627230513

321354BV00004B/211/P